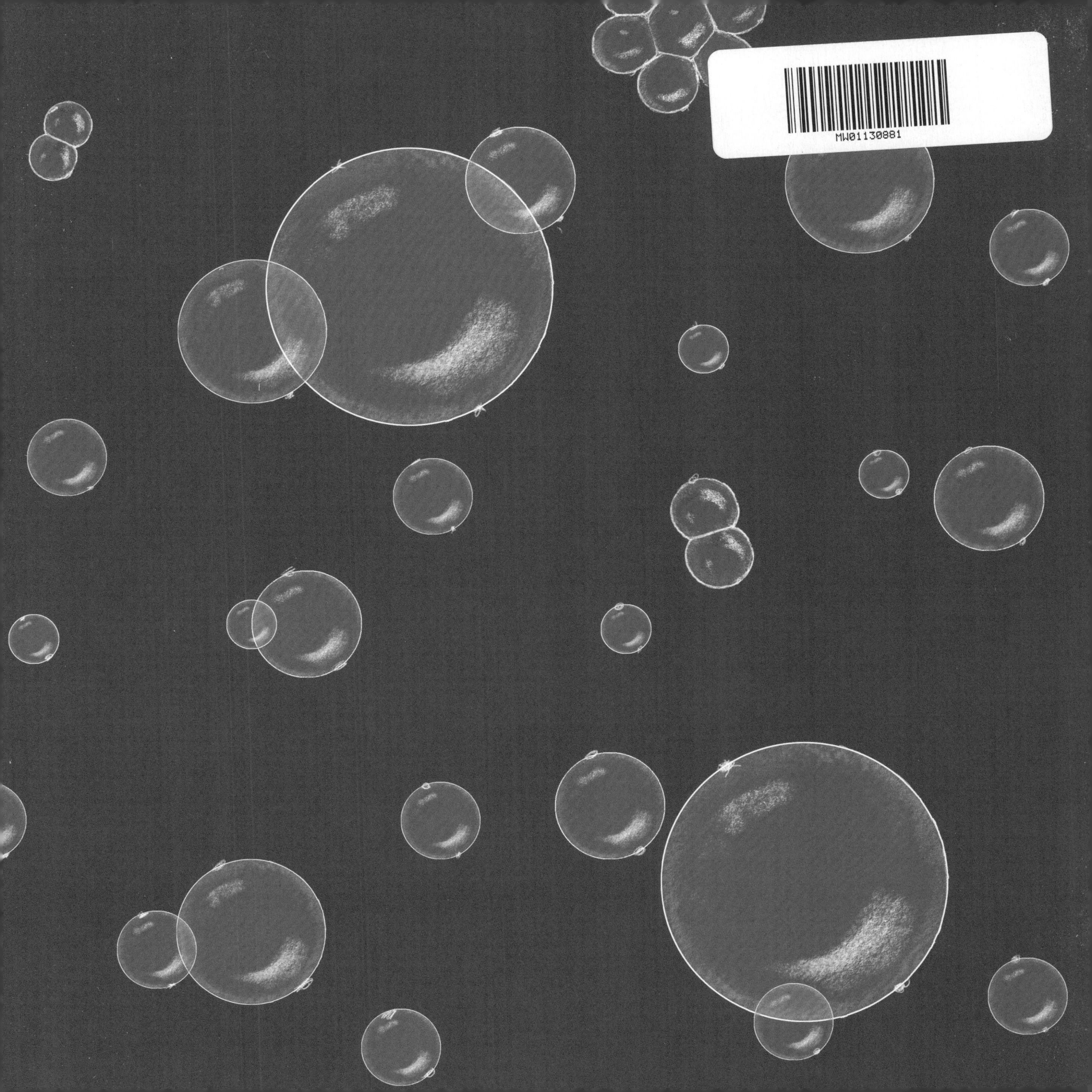

THIS BOOK BUBBLES OVER

For Perran, Harbor, and Sabina, who make me—and one another—bubble over with happiness

—N. N.

For my grandkids, who bubble over with energy and have me struggling to keep up

—R. M.

Published by
PEACHTREE PUBLISHING COMPANY INC.
1700 Chattahoochee Avenue
Atlanta, Georgia 30318-2112
PeachtreeBooks.com

Edited by Kathy Landwehr
Design and composition by Lily Steele
The illustrations were digitally rendered.

Printed and bound in November 2024 at C&C Offset, Shenzhen, China.
10 9 8 7 6 5 4 3 2 1
First Edition
ISBN: 978-1-68263-731-9

Cataloging-in-Publication Data is available from the Library of Congress.

EU Authorized Representative: HackettFlynn Ltd, 36 Cloch Choirneal, Balrothery, Co. Dublin, K32 C942, Ireland. EU@walkerpublishinggroup.com

THIS BOOK BUBBLES OVER

From the Ocean to Mars and Everywhere In Between

Written by
Nora Nickum

Illustrated by
Robert Meganck

PEACHTREE
ATLANTA

What is a bubble?

A puff of air,
a swirl of gas,

temporarily trapped
in something else.

Perhaps just moments
away from popping and
disappearing forever.

A bubble might look flimsy
and insubstantial.

But there's more to it than that.

A bubble can be noisy.
Delicious. Useful. Dangerous.

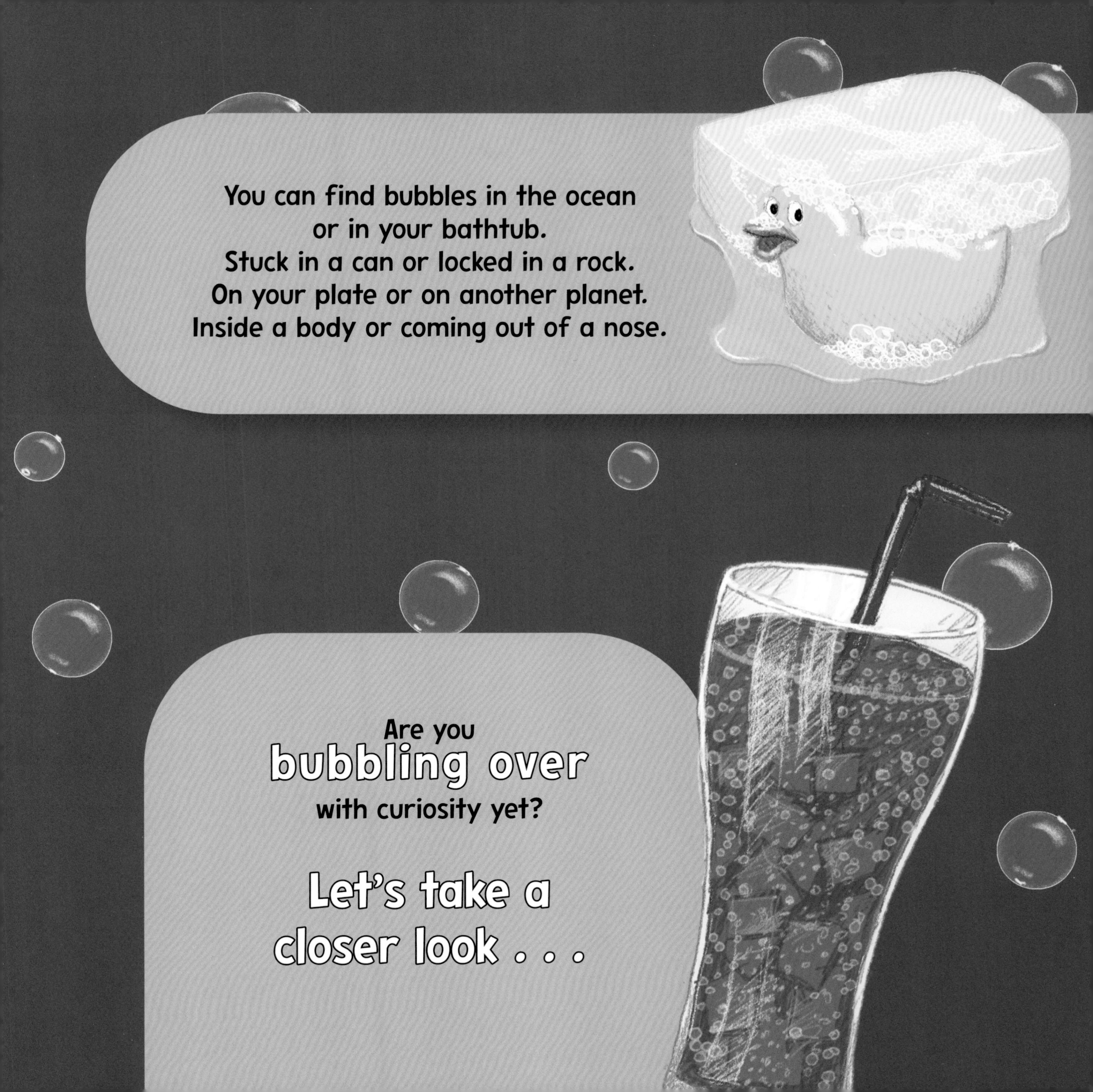

You can find bubbles in the ocean
or in your bathtub.
Stuck in a can or locked in a rock.
On your plate or on another planet.
Inside a body or coming out of a nose.

Are you
bubbling over
with curiosity yet?

Let's take a closer look . . .

A bubble can be solitary . . .

The material used to make bubble gum is more flexible than what is used for regular chewing gum. That means it can stretch farther as you blow air into it. One Guinness World Record holder used three pieces of bubblegum to blow a bubble 20 inches (51 centimeters) in diameter. That's big enough to hold your whole head!

. . . or one of many.

How do you make whipped cream? Grab a mixer, a bowl, and a carton of whipping cream, then get to work. Whisking introduces air into the cream, producing zillions of bubbles. They're so small that you might not see them, but they're making the cream gloriously light. The fat in the cream gathers around the bubbles and keeps them from popping. What about a can of whipped cream from the store? It contains a less appetizing soup of liquid cream and dissolved gas. When you push the nozzle, bubbles of nitrous oxide form within the cream, and it becomes fluffy and scrumptious as it's pushed out of the can.

A bubble can last a long time . . .

Bubble Wrap can cushion a fragile gift, so it won't break on its trip through the mail system. But how do the bubbles get into the wrap in the first place? A machine melts plastic beads and squeezes them through a slot. The resulting sheet is fed into a metal roller, which sucks the plastic into a bunch of round holes, creating pockets. A second plastic sheet is then placed on top and sealed, trapping air in between. The bubbles in Bubble Wrap last a long time—at least until you're ready to make some noise.

. . . or pop right away.

When you blow a bubble, you create a slim sandwich composed of layers of soap, water, and more soap. It's fragile and might pop if it bumps into something. But even if it doesn't, it won't last very long. Some of the water begins evaporating almost immediately. Gravity causes the rest of it drain to the bottom of the bubble. Pretty soon your bubble will be too thin at the top or too dry to hold together. Pop! Want to make the next one last longer? Blow it outside on a cooler day or add corn syrup to the mix. Both of those will slow evaporation.

A bubble may be shaped like a sphere . . .

Blow another bubble and watch it break free from your wand and float into the air. It will form a sphere, the most compact shape that can contain the air inside. Other shapes would require the surface to stretch more, and unevenly, to hold the same amount of air. It's not really possible to blow a bubble shaped like a cube or a pyramid, but if you did, it would be more fragile and thus more likely to pop. And why does the bubble float? The air inside the bubble is your warm breath, which is lighter than the cooler air outside it.

. . . or a hexagon.

When a bunch of similarly sized bubbles meet, their walls join up and their shapes change. Spheres no longer, they each become more like hexagons. That's the same efficient shape made by bees building a honeycomb.

A bubble can be tiny . . .

The star-nosed mole, found in wetlands in North America, searches for food by rapidly exhaling through its nostrils underwater. Each breath forms a bubble—as many as five to ten bubbles every second. When one bumps into something, it picks up the smell of that object. The mole inhales the bubble—and the scents it has captured—before it can float away from its nose. All it takes is a sniff to determine whether it has come across a tasty meal, like a fish, frog, or worm, or something less scrumptious, like a rock. Scientists used to think mammals couldn't smell underwater, but thanks to bubbles, apparently these moles can!

. . . or **enormous.**

When an undersea volcano erupts, it can produce a bubble bigger than a stadium. Hot magma, full of gas, flows out of the vent on top. Cold seawater chills the magma, which forms a hard lava cover over the vent. As the shallow volcano continues to erupt, more gas pushes on that cover, creating pressure. The growing bubble rises above sea level and when the cover finally breaks, the gases are released into the air.

Bubbles can be

inside a liquid . . .

When you boil water, two kinds of bubbles develop. Cold water contains gases (nitrogen, oxygen, and carbon dioxide) absorbed from the surrounding air in the room. As the water warms, those gases form small bubbles on the bottom of the pot, which rise to the surface and pop. Larger bubbles form later as the temperature increases and some of the water becomes steam. Now's a good time to add spaghetti. These larger bubbles will help you cook your food; their movement spreads heat throughout the water, keeping the noodles from sticking together and ensuring they cook evenly.

. . . or a solid.

Bread dough is made from flour, living fungi called yeast, and other ingredients. Kneading these ingredients together works air from the room into the mixture, creating bubbles. Then, the yeast digests sugar found in the flour, producing carbon dioxide. This gas makes the bubbles expand and the dough rise. The bubbles get even bigger in the oven, where heat turns the water in the dough into steam. Next time you eat a sandwich, check out the pockets left behind by the bubbles. Without them, you'd be trying to gnaw a dense, cracker-like brick, instead of savoring something soft and light.

Bubbles can keep an animal **warm . . .**

Many surfers wear wetsuits made of neoprene foam, which contains microscopic bubbles of nitrogen gas. Heat can't pass through nitrogen as easily as it can through water or a solid, so those air bubbles prevent the surfer's body heat from being quickly lost to the cold seawater. That means there'll be time to catch a lot more waves. Surf's up!

. . . or cool it down.

The short-beaked echidna lives in very hot parts of Australia, but it doesn't pant like a dog to cool off, and it can't sweat like a human since its skin is covered in a blanket of spines. How does it beat the heat? By blowing bubbles from its pointy nose. Those bubbles burst over the tip of the nose, covering it with wet snot. As that mucus evaporates, the echidna's body heat dissipates. Cooled by this process, the blood under the skin of its nose travels throughout its body, reducing the temperature everywhere it goes.

The violet snail spends its whole life in the open ocean, but it can't swim. It must make its own flotation device—and it can't exactly hammer together some driftwood. What else is available? Bubbles and snot. The snail stirs up the water with its foot, creating bubbles, and adds mucus to glue those bubbles together into a raft. Then it attaches itself to the underside and sails away.

A young dolphin blows air through its blowhole, creating bubbles and bubble rings to play with. It may chase them, swim through them, or even bite them. While bubble games look like great fun, the exercise also helps the dolphin get stronger and practice moves that will come in handy later for catching food or escaping from danger.

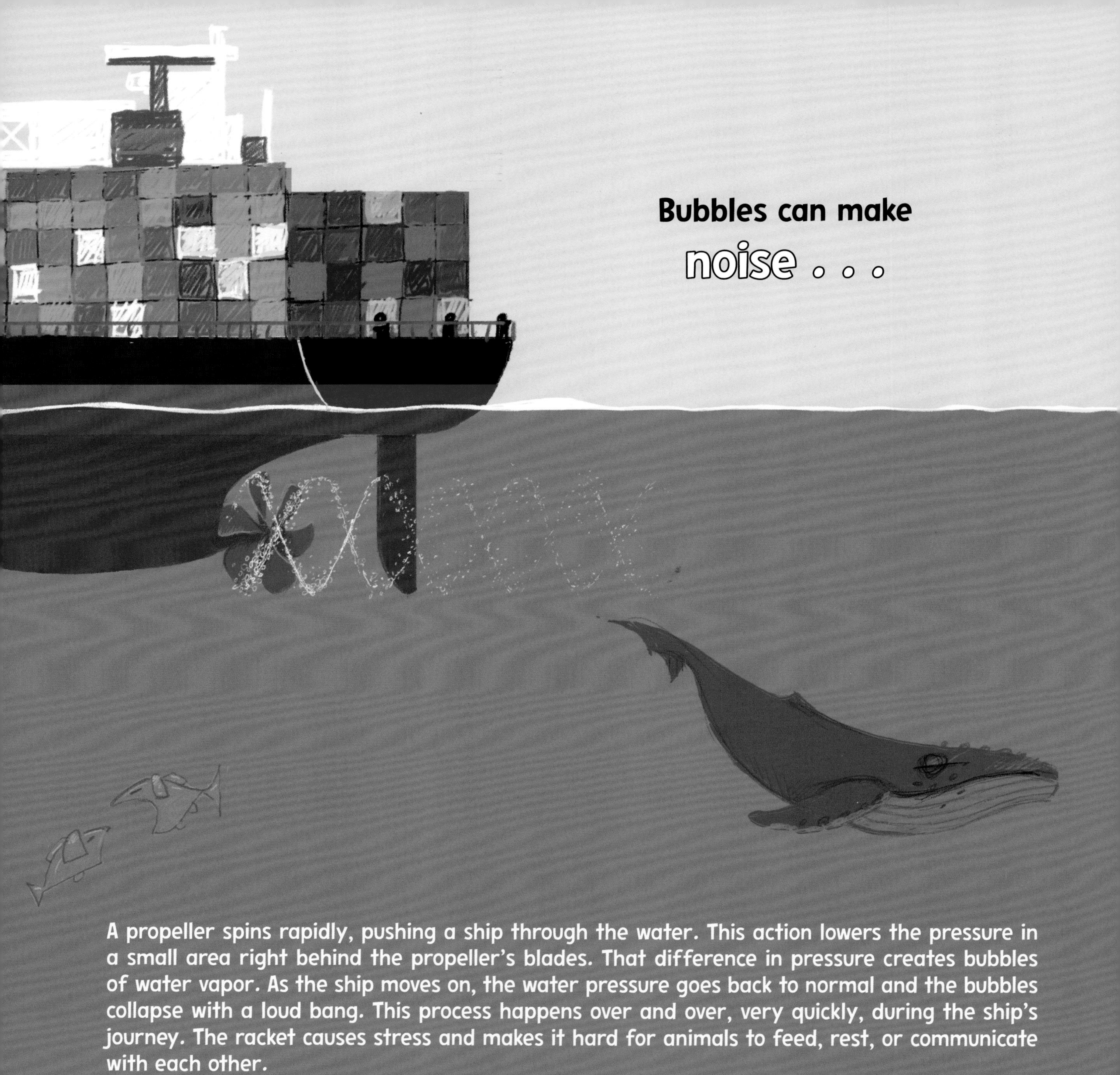

Bubbles can make noise . . .

A propeller spins rapidly, pushing a ship through the water. This action lowers the pressure in a small area right behind the propeller's blades. That difference in pressure creates bubbles of water vapor. As the ship moves on, the water pressure goes back to normal and the bubbles collapse with a loud bang. This process happens over and over, very quickly, during the ship's journey. The racket causes stress and makes it hard for animals to feed, rest, or communicate with each other.

. . . or create quiet.

Workers use huge hammers to build things like wind farm foundations in the ocean. The pounding can be loud enough to scare porpoises and other animals away from important feeding areas or damage their hearing. Fortunately, people are working to make construction sites quieter. A tube, punctured with lots of holes, can be placed on the seabed, surrounding the site. A machine blows air into the tube, which then escapes through the holes. The rising bubbles make a curtain to muffle the hammering noise. Some sounds reflect off the bubble curtain, so they stay within the construction site, and others are absorbed by it. Less noise gets out to harm animals.

Bubbles can make people sick . . .

During a spacewalk, an astronaut wears a spacesuit that provides oxygen for her to breathe, as well as protection from the radiation and extreme temperatures in space. The air pressure inside the suit is low to make it easier for her to move around. But putting on the suit causes the pressure surrounding her body to change suddenly. When the pressure drops, nitrogen gas that is usually dissolved in the blood can form bubbles instead. Those bubbles block the flow of blood through the body, damage the vessels, and cause pain. This is called decompression sickness. To prevent it, an astronaut breathes in pure oxygen for a few hours before she puts on her suit. As she breathes out, nitrogen leaves her body, so it's not around to bubble up later. A scuba diver can also experience decompression sickness but can avoid it by coming up to the surface slowly after finishing a dive.

. . . or help them get **better.**

Medicine given as pills or shots is distributed through the whole human body. That's fine if many parts of the body need the medicine—like if you have the flu—or if it doesn't have a lot of side effects. But it's not so good if a patient needs strong medicine in just one spot, like a tumor, and wants to avoid that medicine causing damage elsewhere in his body. Scientists are testing out teeny-tiny bubbles to see if they can deliver medicine to a specific location. Doctors inject the bubbles, which contain medication, into the blood in a patient's vein. Then they can watch the bubbles travel through a patient's body on ultrasound images—which use echoing sound waves—and maneuver them to the right spot with magnets. Once the bubbles are in position, the power on the ultrasound can be turned up to make them jiggle and pop, releasing the medicine.

Bubbles can be **protective** . . .

Some fires can't be put out with water—like one involving gasoline, which will spread on top of the water and continue to burn. Instead, firefighters use a special kind of foam, which acts like a blanket of little bubbles that covers the burning gasoline. Those bubbles stop oxygen from getting in to feed the fire and vapors from getting out. If the oxygen and vapors can't mix, the fire will stop burning. Some of the bubbles may pop, leaving holes in that blanket. In that case, there's a second line of defense: chemicals in the foam also leave a thin film on the gasoline, which keeps those vapors from mixing with oxygen while helping to reconnect the foam blanket. These are toxic and they remain after the fire is out, polluting the environment for a long time. Engineers are trying to invent safer firefighting foams that will still use the power of bubbles.

. . . or dangerous.

When humpback whales gather at their feeding grounds, they may use bubbles to create something that'll put fear into any fish: a net. One whale swims in an upward spiral around a school of fish, blowing bubbles from its blowhole that rise up in a shape like a net. Others surround the little fish and herd them farther into the bubble net. Then all the whales swoop up and eat as many of those trapped fish as they can.

Bubbles can change the **weather** . . .

As waves form and crash in the windy open ocean, they trap air and push it down into the water, where it forms tons of bubbles. When the bubbles rise to the surface and burst, they spray seawater up in the air. The water from that spray evaporates, leaving behind specks of salt and particles from ocean life like phytoplankton, which can all be blown higher into the air by the wind. They may be tiny, but they are important—because to make clouds, water vapor in the atmosphere needs something like those traces of salt to attach to. Without bubbles bursting and leaving particles behind, clouds wouldn't form over the ocean and rain wouldn't follow.

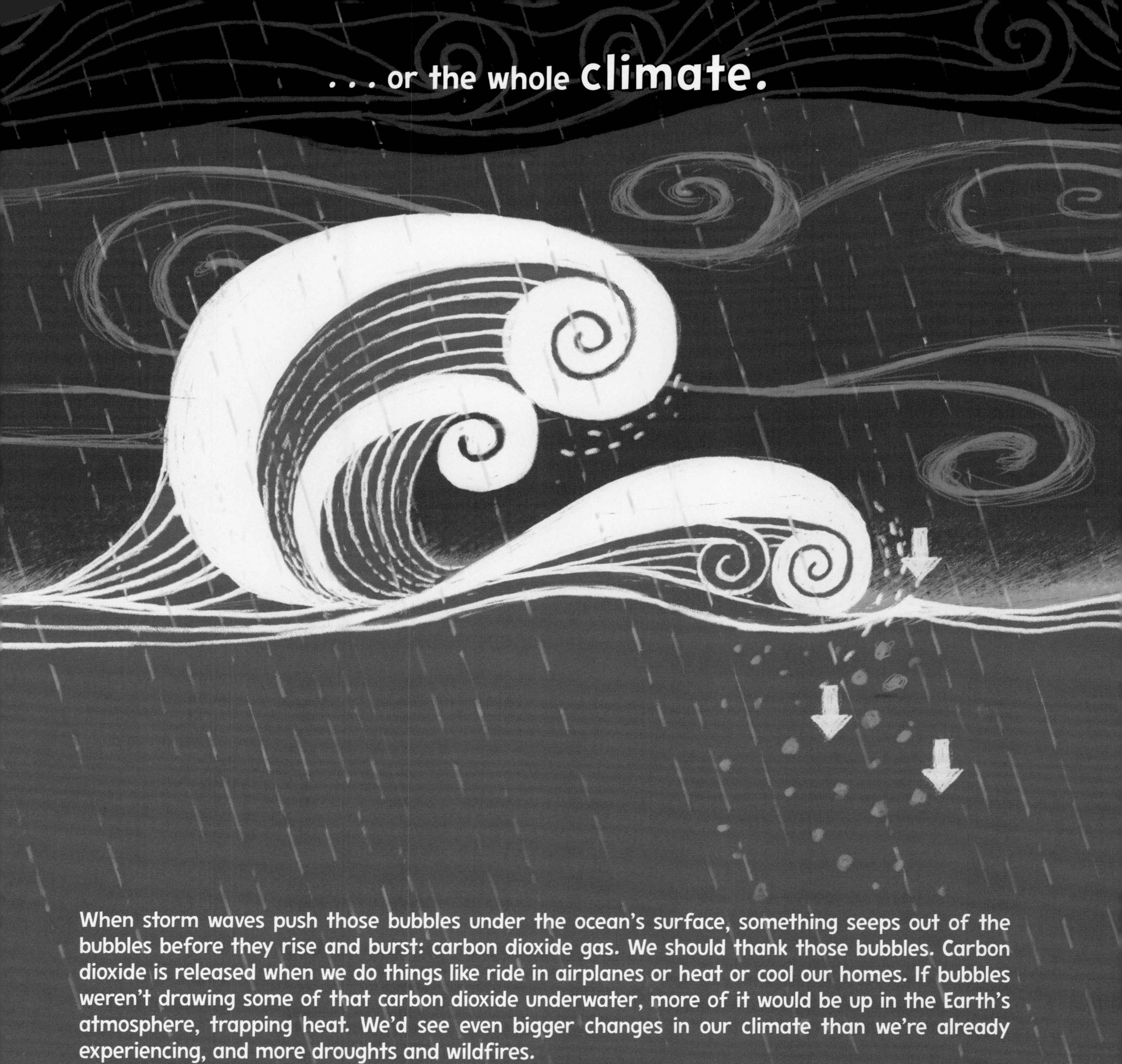

. . . or the whole climate.

When storm waves push those bubbles under the ocean's surface, something seeps out of the bubbles before they rise and burst: carbon dioxide gas. We should thank those bubbles. Carbon dioxide is released when we do things like ride in airplanes or heat or cool our homes. If bubbles weren't drawing some of that carbon dioxide underwater, more of it would be up in the Earth's atmosphere, trapping heat. We'd see even bigger changes in our climate than we're already experiencing, and more droughts and wildfires.

Bubbles can reveal the history of the Earth . . .

One of the reasons we know our climate is changing, and why, is because of information scientists have gathered from bubbles deep in the ice in Antarctica and Greenland. Scientists use rotating or heated pipes to drill into that ice—sometimes just 60 feet (18 meters) deep, sometimes as much as 2 miles (3 kilometers)—to gather samples that contain tiny bubbles of ancient air. These bubbles were trapped hundreds of thousands of years ago as layers of snow built up. Scientists melt or crush the ice in a vacuum (something fancier than what you have at home) to remove the air. By measuring the concentrations of different gases in that air, they can find out what the atmosphere used to be like. Thanks to these bubbles, we know the amount of carbon dioxide in the atmosphere today is unusually high compared to what it has been historically.

. . . or the history of life on Mars.

Ancient bubbles in basalt rocks found on Mars might someday reveal evidence that something has lived there. Long ago, lava, containing gas bubbles, flowed from volcanoes on the planet. When the lava cooled and hardened into rock, some of the bubbles were trapped forever. Those hardened little holes could have become safe places for creatures called microbes (such as bacteria) to live, even if nothing could survive on the planet's now-waterless surface. That's because the bubble remnants offer not only living rooms, but also places where water and nutrients could move through the rock. And unlike some other kinds of rock, basalt is strong so it wouldn't collapse in on these tiny beings. Testing has shown bacteria lived in basalt bubbles on Earth more than 350 million years ago. It's possible Martian bacteria have done the same.

ENGLISH IS FULL OF BUBBLES

People have found imaginative ways to use the word "bubble" to describe concepts and feelings, not just physical things you could pop with a pin.

If you're **bubbling over** with happiness, you're so full of joy you can't contain it, just like oatmeal bubbles over the side of the pot on a hot stove. You can bubble over with other strong emotions, too, like anger or excitement.

If someone is **living in a bubble**, it's as if they are separated from the real world by their assumptions, beliefs, or imagination. They ignore ideas that are different from theirs and stick with people who think like they do.

If you **burst someone's bubble**, you make them realize what they believed isn't true, or what they hoped for won't happen, or that something isn't as great as they thought. They feel disappointed, like you might if someone popped your huge soap bubble before you had a photo for the world record competition. Nobody quite knows where the saying originated, but it appears to have been in use as many as 100 years ago.

Someone with a **bubbly personality** is lively and enthusiastic, seeming to overflow with cheerfulness even in situations where others might be bored or tired.

Look around you today and see what bubbles you can spot. You might find some in the bathroom or the kitchen, in science class or in the mailbox, in the pool or at the park. Big or small, useful or fun, noisy or quiet, poppable or more permanent, they all deserve a nod of appreciation before they burst. And then make some of your own!

DIVE DEEPER INTO BUBBLE SCIENCE

Water molecules have high **surface tension.** They want to stick close to the other molecules below and beside them. That stickiness stops the water molecules from stretching up and forming bubbles away from the other water molecules. If a bubble does temporarily form, like when you run more water into the bath, it'll pop right away as the water molecules involved hurry to get close to the others underneath.

When you add soap to the bathwater, it spaces out the water molecules and weakens their pull on each other. That lowers the surface tension. The soap and water mixture can stretch around air to make a longer-lasting bubble.

Foams—like milkshakes and shaving cream—are light and frothy thanks to a lot of small bubbles. Bubbles made purely of water will quickly pop, so if you want to make a foam that'll stick around long enough to eat or shave, you'll need to add a **surfactant.** That's an ingredient like fat or soap that sticks to the water's surface and helps the bubbles last longer. How do the little bubbles form in the first place? Air can be brought into a liquid by shaking, stirring, or pouring it, or by adding chemicals that make gas. If you're whipping something up in the kitchen, you'll usually find that stirring faster creates smaller bubbles and helps the mass of foam grow bigger.

AUTHOR'S NOTE

When you write a book that includes all sorts of different bubbles—found not only in the ocean, but also on Mars and everywhere in between—you have to gather information from a lot of different sources. And then you have to keep looking for *more* sources, because it's important to double- and triple-check any specific facts you want to include in your book.

Wherever possible, I used recent articles from scientific journals so the information would be both up-to-date and accurate. I also checked websites and publications by universities, government agencies, and museums. A few examples:

- I read about echidna snot bubbles and underwater volcano bubbles in the journal *Science* and learned more about medicine bubbles in *Cell & Bioscience*.
- I turned to the National Science Foundation and the *Annual Review of Fluid Mechanics* to understand how ocean bubbles contribute to cloud formation, and to NASA and the Harvard Medical School to investigate decompression sickness.
- *Astronomy & Geophysics* and NASA described bubbles on Mars, and the American Museum of Natural History and National Wildlife Federation websites revealed fascinating facts about star-nosed moles.

I bubbled over with delight whenever I found a great, reliable source and learned something new. And then I found an actual bubble expert! Dr. Helen Czerski at University College London researches the physics of ocean bubbles. She reviewed what I'd written and recommended a few edits for accuracy. I'm grateful for her time and expertise; any remaining errors are mine.

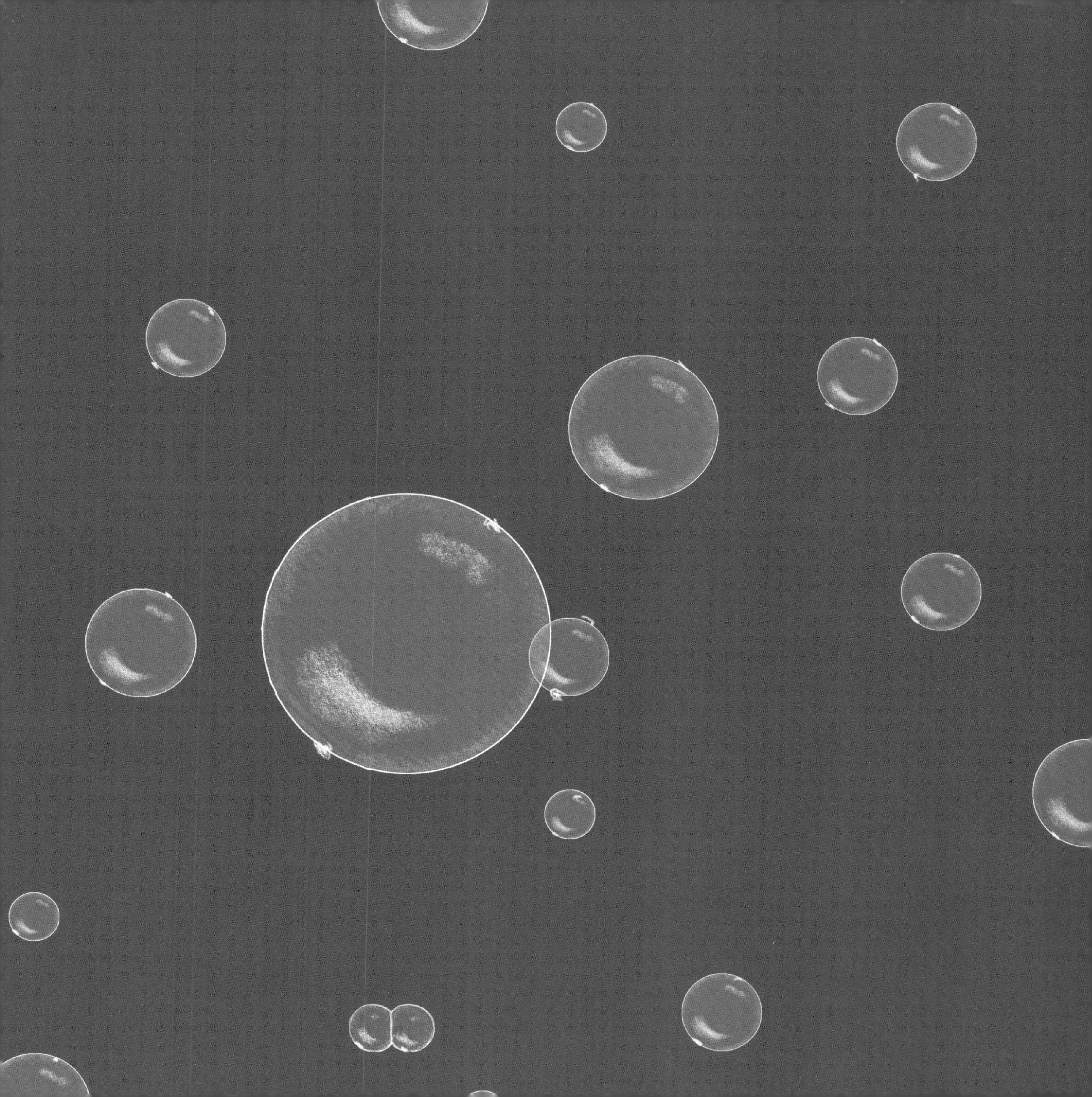

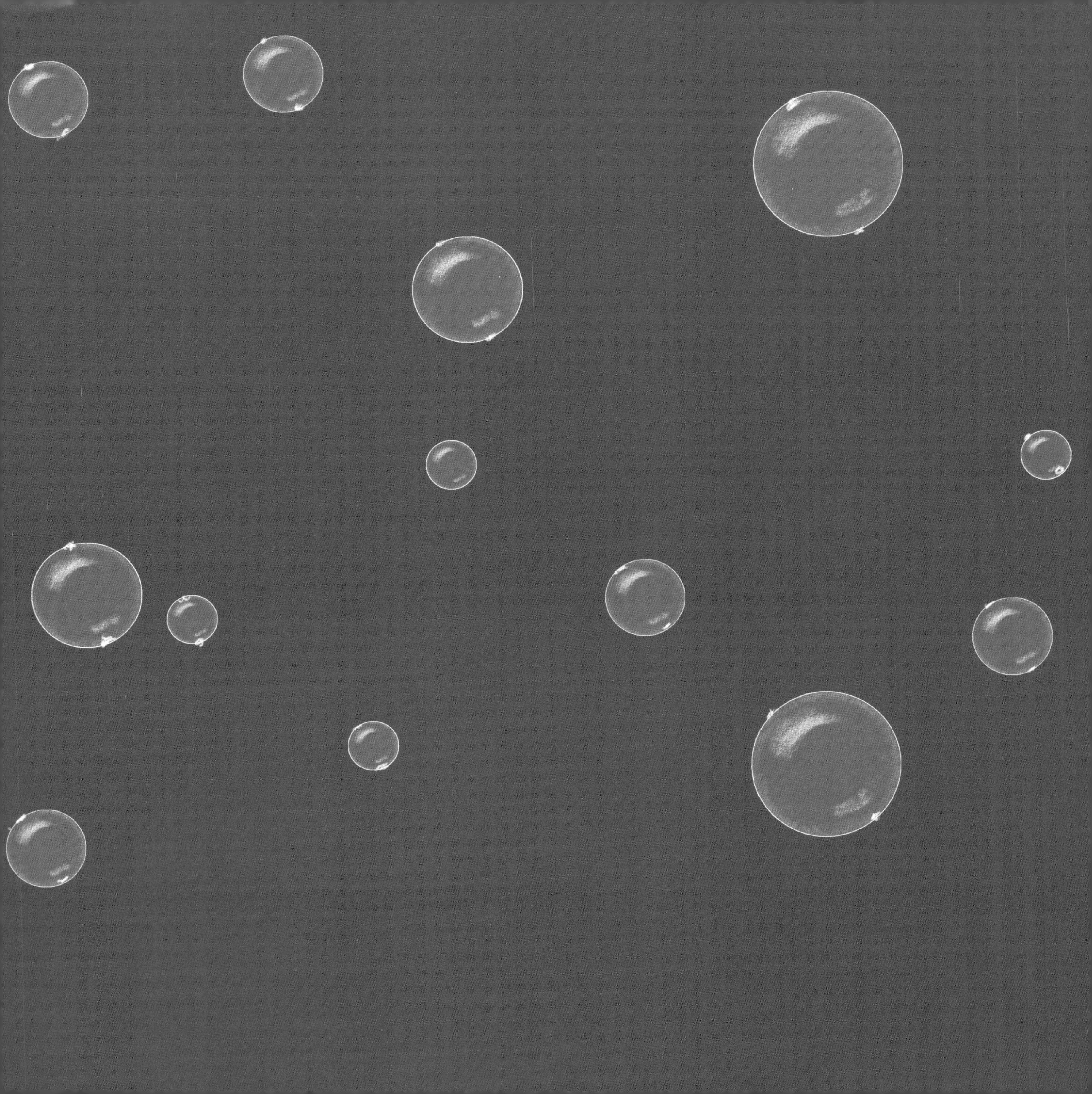